Anthropocene
The age of man

By David Millett

CONTENTS

INTRODUCTION

Over the billions of years that the Earth has existed natural forces have enabled a record of its history to be stored in its rocks and sediments. Events such as volcanism, asteroid and comet impacts, climate change, and constant erosion by wind, water, and sunlight are all captured in this history book of deposits, layer after layer.

During the Earth's long history life has left a record too. Life's history is not just captured in the form of fossils and micro fossils, but also in the global effect it has had on the Earth. In this layered geological time-capsule we find the fossils of what are thought to be the first users of photosynthesis, filamentous photosynthetic organisms dated to 3.4 billion years old. We can also see the chemical signature of the oxygen they produced in the rocks and sedimentary layers.

By studying this layered time-capsule of rocks and sediments since the Earth formed we are able to read the history of our planet just as if it were written down in a history book. We can tell when Earth formed 4.6 billion years ago by radiometric dating of meteorite material. Comparing these findings with the age of the oldest-known terrestrial and lunar rocks, we can estimate when the Moon formed from a catastrophic collision between the new Earth and another planet sized body some 4.5 billion years ago. We know from the rocks and sedimentary layers when heavy astronomical bombardment of the Earth's surface stopped and when the first life formed some four billion years ago. Our sedimentary history book tells us when the Earth's surface first became entirely frozen 2.3 billion years ago, when life rapidly changed from mostly individual colonies of cells into more complex animals 530 million years ago, when the first vertebrate land animals arrived 380 million years ago, when the dinosaurs first began their long reign some 230 million years ago, and our layered time-capsule tells us when the first Hominids (our ancestors) evolved two million years ago.

Given our new insights into the Earth's past we are at a point where we can predict its possible future. Nothing is ever certain, but given the data we have, reasonable extrapolations and predictions are possible.

WHAT IS THE ANTHROPOCENE?

GEOLOGICAL TIME PERIODS

Geochronology and biostratigraphy are two systems used by geologists to determine the age of rocks, fossils, and sediments. Over the years they have used these and other methods to classify the Earth's history into periods of time. We are currently living in the Cenozoic Era. There were nine eras before the current one stretching back to when the Earth first formed.

Within the current era we reside in the Quaternary Period, which is the most recent of three periods that make up the Cenozoic Era. The Quaternary Period is further subdivided into the three epochs: Pleistocene, Holocene, and the Anthropocene.

THE ANTHROPOCENE EPOCH

Until recently geologists thought we still lived in the Holocene epoch, which began around 12,000 years ago. In 2008, a proposal was presented to the Stratigraphy Commission of the Geological Society of London to make the Anthropocene the current epoch, and to make the Anthropocene part of the formal system of geological measurement. Steps arc being taken by independent working groups of scientists from various geological societies to determine whether the Anthropocene will be formally accepted into the

Geological Time Scale, but meanwhile many scientists think that we now reside in the Anthropocene epoch.

ANTHROPOCENE INDICATORS

There is growing consensus among most scientists that human activity has left and continues to leave a permanent geological record, and that this record marks the beginning of the Anthropocene: The Age of Man.

It is probable that the Anthropocene epoch started at the beginning of the industrial revolution some 260 years ago when James Watt invented the steam engine. The steam engine was first developed to help extract coal from coal mines. This technology enabled people to get at more and more coal. Then people got the brilliant idea to put a steam engine on a vehicle and put these vehicles on tracks. So began our journey to today.

The evidence is mounting that the Anthropocene epoch started with the industrial revolution. Before this time the geological strata records little effect of humans on the Earth's eco systems.

The Holocene/Anthropocene epoch boundary can be seen from the effects measured in the geological record today by looking for signs of the human global nitrogen cycle, terraforming, agriculture, climate change, ocean acidification, and species extinctions and concentration.

NITROGEN CYCLE

Human activity has affected the natural nitrogen cycle. For eons, nitrogen gas was converted to nitrogen compounds (nitrates) by bacteria in the soil. Plants then used these nitrates to make proteins. Back in the 1930s humans figured out how to extract nitrogen out of the atmosphere and turn it into ammonia to be used as fertilizers (known as the Haber-Bosch process). After World War II we got into nitrogen fertilizer production in a big way. Over a period of only a few decades, we went from a planet where nitrogen was in short demand to a planet that is literally awash in nitrogen.

The burning of gasoline, natural gas, and coal with atmospheric nitrogen produces nitrogen oxides as a byproduct too. The two contributions of human derived nitrogen from artificial fertilizers and fossil fuel combustions have now created a global nitrogen cycle, which is completely due to the human species. We now create more biologically utilizable nitrogen than the sum total of all natural fixation processes associated with all the nitrogen fixing microbes in the ocean and in the soil.

This human derived nitrogen cycle is a defining characteristic of the Anthropocene and clearly separate this epoch from the Holocene. This increase in nitrogen leaves distinguishable traces in the crust of the Earth. We find evidence of nitrogen in the geological record in high mountain lake ecosystems as well as in lakes in the high Arctic. Extracting sedimentary rock core samples from the bottom of the lakes in these locations can tell us about changes in the environment over time. Sediment cores are highly organized stratigraphic archives of environmental change. Analysis of the differences in both the amount and the isotopic character of the nitrogen preserved in the tops of these sediment cores is compared to stratigraphic markers just a few centimeters beneath the surface. These comparisons demonstrate a greater concentration of nitrogen in these widespread landscapes.

Mass spectrometry enables scientists to very carefully measure the two stable isotopes of nitrogen, N-15 and N-14 within the organic matter in these sedimentary samples. The study provides a clear fingerprint for human derived nitrogen (N-15). Both naturally created and manmade nitrogen chemical signatures are archived in the geological record as it forms. Should future generations of geologists look they will be able to identify these chemical signatures. The presence of the manmade isotope of nitrogen is demonstrable evidence of human influence at a planetary scale.

Also this human modified global nitrogen cycle is causing huge and very visible problems. Human created reactive nitrogen is washing off the land into the oceans creating coastal dead zones from the Baltic Sea to the Gulf of Mexico. When it enters the atmosphere it becomes a powerful greenhouse gas adding to the global warming issue.

The human nitrogen cycle is a clear boundary marker for separating the period of relative climate stability of the last 10,000 years, the Holocene, from the current Anthropocene epoch.

TERRAFORMING

Human mining (or terraforming) is the next defining characteristic marking the Anthropocene. Humans (rather than ice, wind, and rivers) are now the largest force in the movement of sediments on our planet. For example, we currently mine eight billion tons of coal per year up from four billion tons in 1980. By 2030 we will reach 13 billion tons of coal mining per year. This will be equivalent to all the sediments that rivers currently move to all coasts. This massive increase in the movement of earth does not include the 13 billion tons of gravel and sand (aggregate), or the two billion tons of hydraulic cement, or the two billion tons of iron ore we currently extract and move.

There are many ways that humans are mobilizing this material on our Earth surface today and we are changing its location and piling it up in ways that exceed natural forces. It is estimated that there are 568,000 abandoned mines in the USA alone and many millions more throughout the world. We are sculpting the Earth's surface, taking material away and leaving surfaces that are not natural. We are piling the sediment up at speeds and rates that we have not seen before.

Compared to the geological past (which usually drags things out over thousands if not millions of years) this current accelerated rate of earth movement has happened over the last 200 years and is something astounding. It is probable that this terraforming will only increase as we continue to build million person cities, every ten days, over the next 87 years.

Human terraforming is also causing massive changes to the natural environment of the planet. We have been building one large dam every day on average for the last 130 years. There are now hundreds of thousands of dams and reservoirs around the world. Each restricts the sediment that would otherwise normally flow to the ocean. The dams are trapping these sediments. This causes fewer of these sediments to reach and fertilize agricultural areas including the rice-bowl-deltas of the world.

By mining water for agriculture, for industrial use, and for human consumption, we are lowering the land four times faster than the ocean level is rising due to global warming. Half a billion people, who now live on these rice-bowl-deltas, are much more susceptible to things like storm surges and tsunami waves.

We no longer allow rivers to flow where they once did. The Yellow River (in Asia), for example, would migrate across its 700 kilometer flood plain changing its course from time-to-time. We no longer want rivers to migrate anymore because that would impact the location of roads, cities, and towns. We now keep rivers fixed by levy and other systems. Because of human influence most of the rivers on the planet are no longer considered natural.

MODERN AGRICULTURE

Another dramatic signature of the Anthropocene is modern agriculture. If you look globally at the production of food and timber, especially food production, about 40 percent of the ice free land surfaces of the Earth are now used for pastures and production of crops. These farmlands influence the wild landscapes surrounding them. Modern agricultural land use affects patterns of bio-diversity, habitat structure, and the cycle of water. Simply put, modern agriculture is the biggest influence humans have ever had on the biosphere.

We can see today habitat modification made to support our ever growing domestic livestock, which is used to feed our ever increasing population. Cattle, sheep and these sorts of species depend on grasses primarily for their diet. And so we have in many cases literally cleared away forests to accommodate our food source's need. Examples such as the deforestation of Europe, North America, and South America are the most prominent. We are continuing to clear forest from around the world to make way for grasslands to feed our livestock in a veracious manner. And in other areas we are destroying forests to plant specific plant species that are productive to us for their feed. We are going to be using a huge amount of the primary productivity of plants to feed all these species.

CLIMATE CHANGE

Our reliance on, and over use of fossil fuels is adding a new chapter to the Earth's history. The evidence points to our dependence on these fuels as adversely affecting the Earth's climate systems.

Does our geological history book record similar past climate change events?

In railway cuttings in the countryside near Leicester, England, scientists discovered a geological record of life and death in a 180 million year old sea floor. In the lower yellowish limestone layers of strata they discovered a rich fossil record of many creatures that lived on this ancient sea floor back then, such as shells and mollusks. Just above this layer are black shales that indicate the sea floor became stagnant and effectively dead. This layer is a record of a sea with no oxygen. All of the fossil life found in this layer lived in the high sunlit waters above and when they died they simply fell onto the stagnant sea floor.

The change from yellowish limestone to the black shales in these layers is something that has been closely studied in the last few years. This is because it seems to mark a fossil global warming event when something of the order of two to four trillion tons of carbon were released into the atmosphere and warmed the Earth by about five to seven degrees centigrade. This warming occurred more at the North and South Poles and less at the Equator, but it warmed the ocean, which caused it to become oxygen depleted. This hyper thermal event happened a little over 180 million years ago. It appears to have caused sea levels to rise and many creatures to become extinct. This geological record is very important to understand and study, as we may be at the beginning of a similar hyper thermal event of our own creation.

Speed of climate change

Some scientists posit that the speed of climate change is in some ways more important than the change itself. It is frightening how quickly climate is changing globally today. It is important to be able to provide quantitative metrics by which to evaluate the nature of this modern warming. Work is being done to determine whether the current change is unusual, or normal and within the range of natural variability.

What scientists are seeing now is a rapidly accumulating volume of secure data that shows that the present warming trend is the steepest it has been in tens of thousands of years. This places current global warming way outside the range of natural variability.

How rapidly can climate change occur safely? For example how rapidly can Greenland and West Antarctica shed ice to the ocean? There are billions of people who are vulnerable to a sea level rise of just one meter. A one meter rise in sea level would render cities like Miami in the USA inoperable. A sea level rise of a meter is the current conservative prediction for the end of this century. We know sea levels are rising now and that they will continue to rise in the future.

It is possible that sea levels will rise not by small amounts but by meters and very quickly, given the level of CO_2, methane, and other greenhouse gases in the atmosphere today. Current levels of atmospheric CO_2 is at 400 parts per million. The geological record shows that the last time they were at these concentrations the sea level was 25 meters higher.

The mid-Pliocene is the most recent time when CO_2 levels were anywhere near the levels they are predicted to reach; this was about four million years ago. Back then there were forests all the way to the Arctic Ocean covering all land masses. There was no Greenland icecap. Sea levels were 25 meters higher. The mid-Pliocene gives us a good example of the world we might be heading to, given our current CO_2 generation.

If we were suddenly transported back to the mid-Pliocene we would find the place acceptable to live in, climate wise. However, most of our cities and infrastructure would be underwater as sea levels would be 25 meters higher, but we could survive the climate of the time. If our climate keeps changing at its current accelerated rate then we face a great challenge. If you have to move New York City in 200 years you can do it. If you have to move it in ten years the task may be impossible.

The boundary between the Paleocene and the Eocene epochs took place about 56 million years ago. During this event, global temperatures increased by about four to eight degrees Celsius worldwide, and a lot of carbon dioxide entered the atmosphere at that time.

We know from the geological record that 4.5 billion tons of carbon entered the Earth's geo-system over at least 5,000 years (known as the Paleocene/ Eocene thermal maximum). This slow release of carbon accumulated and ultimately changed the climate enough to transform the Paleocene into the Eocene epoch.

In the Anthropocene epoch, current fossil fuel burning has so far put less total carbon into the atmosphere and oceans, but the rate of carbon release is happening much faster. The carbon release in the Anthropocene is happening over just a few centuries.

The Paleocene/Eocene thermal maximum, has been cited as a naturally occurring analog for what humans are doing today. It is thought that the carbon emission rates at that time were maybe half or perhaps one gigaton per year. Today humanity is emitting around ten gigatons of carbon into the atmosphere each year. So what we are doing today is 10 or 20 times more extreme than what happened in the

Paleocene. The Paleocene event caused the biggest extinction of creatures that lived on the sea floor in the last 60 million years. So what we are doing now is really unusual geologically. You really have to go to the big mass-extinction events in Earth history to find anything comparable.

MASS-EXTINCTIONS

There are now over seven billion people living on our planet, and this number is rising every day. We have converted almost half the world's wild habitats into farmland to feed ourselves and our livestock. Two hundred years of industrial and urban development has changed the global climate and disrupted ecosystems in multiple ways. Add in the impact of our hunting of animals and the species of the Earth are under a level of threat unparalleled in tens of millions of years. Our activities are a clear marker for a new geological age. There is no doubt that humans are a geological force and we live in a new time period the likes of which the Earth has never seen. Whether our impact will get to the magnitude of the famous mass-extinctions of the past is unclear, but it is clear we will be a marker in the geological history book of the planet.

But what are the grounds for such dramatic assertions. Mass-extinctions and major changes of the creatures on the planet have happened before in the Earth's deep past. Geologists use fossil evidence of these past changes in biodiversity to divide up Earth history into different periods and epochs. They have names like the Triassic, Jurassic, and Cretaceous. Most of the geological epochs are in fact defined on the species that lived at the time. So we see different assemblages of species marking each of these different geological time intervals. Sometimes they are marked by extinctions and in extreme cases, mass-extinction events. As we have seen the name for the epoch of the past 10,000 years is the Holocene, but the Anthropocene seems to be a boundary of species extinction too.

What would the fossil record of the future look like? Would it really look different from what we now know as the Holocene?

The end Triassic mass-extinction is found in the geological record at 201 million years ago. The event was triggered by a series of gigantic volcanic eruptions that spewed out vast amounts of carbon dioxide gas. This greenhouse gas drastically changed the climate and transformed the world's ecosystems wiping out many species. More than 100 million years later a gigantic asteroid crashed into the Earth and caused another more famous mass-extinction of the dinosaurs, along with most other land and marine species. They were killed by the searing fallout and climate change that followed the collision. This kind of thing has happened five times in the Earth's history. Today, we the architects of the Anthropocene, are creating a sixth mass-extinction event in Earth's history. What has happened over the reign of the human species is comparative to what happened when the asteroid hit and wiped out the dinosaurs, except now we are the asteroid.

It is clear our activity is causing other species to become extinct. We are overharvesting fish in an unsustainable way by catching more fish that the oceans can produce. The way that we fish, such as bottom trawling, can also cause terrible ecosystem damage. This excessive fishing means that we are threatening our own food supply.

There is no doubt that abrupt climate change in past times has played an enormous role in the extinction of life on Earth. However, some research has discovered that human actions has played a direct role in the extinction of animals, plants, and other organisms and may go as far back as the late Pleistocene, over 12,000 years ago. While previous mass extinctions were due to natural environmental causes, research shows that wherever on Earth humans have migrated, other species have gone extinct. Human population growth, most prominently in the past two centuries, is regarded as one of the underlying causes of what is known as the Holocene extinction event.

Humans have contributed to this mass extinction in three main ways: the increase of global concentration of greenhouse gases, affecting the global climate; oceanic devastation, such as through overfishing and contamination; and the modification and destruction of vast tracts of land and river systems around the world to meet solely human-centered needs. Other related human causes of the extinction event include deforestation, hunting, pollution, the introduction in various regions of non-native species, and the widespread transmission of infectious diseases. At present, the rate of extinction of species is estimated at 100 to 1,000 times higher than the historically typical rate of extinction. It is also the only known mass extinction of plants.

If we do not make a conscious effort to reduce our current loss rates it is inevitable that we will create a situation that parallels the extinction of the dinosaurs. This could certainly happen within a couple of centuries and very possibly much sooner than that, given our current rate of species obliteration. We also need to be aware that extinction trajectories tend to hit a threshold where all of a sudden losses accelerate very rapidly.

Past mass-extinction events are not just marked by spikes in the CO_2 levels found in our geological history book, we also inevitably discover other clues in the strata such as a reduction in species diversification.

The sudden explosion of human population along with our domestic livestock is concentrating current vertebrate species from approximately 200 or 300 classes to a much less diverse population. Humans, make up something like a third of the total weight of all backboned animals that live on land today. Most of the rest of this living mass of life on land is now taken up by all of our domestic livestock, less than ten percent of this matter is made up of all other wild creatures. We have pushed wild creatures to one side in favor of ourselves and our livestock.

In the future geologists should easily find this concentration in land-based vertebrate species in the geological record. There will be bones of cows, pigs, sheep, and goats scattered all through the strata. Human bones, of course, will be there as well and all the plants that we and our domestic animals eat.

We have replaced a lot of primary forest in savannah and jungle with agricultural land where we grow wheat, maize, and barley. This concentration of plant species will also be visible in our geological history book by the pollen they give off.

Our ecological takeover of the Earth is causing creatures to become extinct, these changes are happening now and in a very short geological moment. However, we are also reshuffling creatures from all around the world via mass species invasions, which we are largely responsible for. The scale of this is unique in the Earth's history. There have been species invasions before usually when continents collided, but never at today's rate and scale. The current world-wide species invasion will reset the Earth's biology and be clearly recorded in our geological history book.

Because of differences in evolutionary pressures the biota of South America is very different from the biota of Australia. Evolution has proceeded somewhat independently on these particular continents because they are so separate. These two continents have been on their own evolutionary trajectory for millions of years. However, because we are now moving plants and animals around the Earth, we are acting as, a human version, of plate tectonics, a homogenizer of continents. As we spread animals and plants around the world, over time we will reduce the number of species that we have on the planet. For example, boats landing on islands in the middle of the Pacific or in the middle of the Atlantic deposit rats. We place cats there to solve our self-induced rat infestation. They in turn cause a major dismembering of the fauna and flora in those islands. This is certainly something that will be detected by future paleontologists.

This invasion of species is truly at a global scale now and it is not restricted to animals. Even in some very remote places like Patagonia, scientists have discovered non-native plants, some of which people deliberately planted. For example, such plants as roses now grow wild all over Patagonia along with Scotch Broom, which is a tremendously invasive plant. If future researchers look at the pollen record from our time they will see a clear mark of these plants spread about the world.

Ocean acidification

Increases in carbon dioxide also cause the world's oceans to become more acidic. Ocean acidification occurs when carbon dioxide dissolves in the water and reacts with it to form a weak acid. This acid accumulates in the water and makes the water more and more acidic over time. For the past 100 years we have seen sea water acidity increase by 30 percent. The rate of acidification today is much, much faster than it has been in the past.

Increased ocean acidification has many issues associated with it, none more so than the extinction of the world's coral reefs. Perhaps one-third of coral reefs are not sustainable due to the acidification that has already happened. If we continue current patterns of carbon emissions there will be no place left in the ocean that can sustain coral reefs by the middle of this century. When these estimations were made the middle of the century seemed far in the future, but the middle of the century is not even four decades away now. We haven't much time left before coral reefs will be extinct.

IS THE ANTHROPOCENE MEASURABLE NOW?

The current geological record says yes, it holds direct evidence of the Anthropocene. The ocean floor is like a huge undisturbed basin. All the plankton that lives at the surface eventually die and when it does it all falls to the ocean floor where it accumulates. All that sediment accumulates layer by layer. Thousands and thousands of planktonic micro-fossils are recorded in this sediment. The animals that made these fossils used the atoms that were in the water at the time they lived. When they died and settled to the ocean floor they created a snapshot of the chemistry of the water they lived in. Every sediment core is basically a recording, from millions of years in the past, of everything that happened in the water column above. These cores can tell us what the sea surface temperatures were and how much CO2 was in the water; the amount of information they contain is phenomenal.

We can tell from this record when a geological age changed from one to another; such as when we went from the last ice age to the current warm Holocene period when human civilization flourished. We can now see in this record a similar change that clearly marks the current epoch: the Anthropocene.

There are sediments in certain regions that accumulate very rapidly, if you analyze the top parts of these sedimentary cores you can sample the micro-fossils held in them and measure their carbon isotope ratio. When this is done you discover that the carbon in these micro-fossils has the isotopic signature of all the fossil fuels that we have been burning for the last 200 years. This carbon was released to the atmosphere and has been absorbed by the ocean. The signature of human activities is clearly observable in these sedimentary cores.

OUR FUTURE

The evidence is clear; we now live in a geological period that is influenced primarily by human activity. Natural forces are no longer the main influencer of the Earth's geology: we are. This Anthropocene epoch is new. No single species has ever had such a profound influence on the Earth except, perhaps, the first oxygen producing creatures. But all these naturally driven processes of the past have no precedence for what is happening today.

An argument could be made that we created the problems and therefore we will be able to solve them. However, the evidence for this argument is slim. If we look at the history of human civilizations we see that there is a strong indication of a repeating cycle of growth and then collapse. Ronald Wright has pointed out in his book: "A Short History of Progress" a clear history of human civilization's growth and then collapse.

For example, the civilization that developed on Easter Island was responsible for the destruction of the island's ecological environment, and ultimately the destruction of the human civilization there itself. In the past Easter Island was green and lush. Its rich volcanic soil supported thick forests of the Chilean wine-palm, which is a tree that can grow as big as an oak. It was not a natural disaster such as a volcanic eruption, drought, or disease that caused the island's eco-system to collapse; it was its human inhabitants that triggered the catastrophe on Easter Island. Why did this happen? One argument is their ideology was the cause. At some point the islanders came up with the idea that they must honor their ancestors by mining, carving, and erecting stone monoliths. Indications are that the ritual started out quite modestly. However, the idea became so powerful and so ubiquitous that in the end every tree on the island was eventually chopped down to implement their ideology. Once the forests were decimated the island's eco-systems collapsed and the human civilization fell back to nothing but a handful of Stone Age inhabitants; the only level of human civilization capable of being supported by the island's devastated environment.

Were the Easter Islanders less intelligent than the rest of the world? Absolutely not, the Easter Islanders were Homo sapiens like you and me. If you doubt this think for a moment about our use and reliance on fossil fuels. At the beginning of the Anthropocene 200 years ago our forebears thought it would be a good idea to build machines that ran on fossil fuels. The idea seemed sound at the time and started modestly, but now we live in a world where the very notion of existing without burning tons and tons of fossil fuels every hour seems unthinkable. The fossil fuel ideology is not so dissimilar to the ideology of building stone monoliths on Easter Island. They have in common that they are both human ideas. The difference today as compared to what happened on Easter Island is only one of scale.

When Easter Island's eco-systems collapsed the failure was limited to the island. If the islanders could build a boat they could sail away from the ill-fated island and start again someplace else. Today, the ideology of fossil fuel use is a worldwide phenomenon. We cannot sail away to any other place. We have one world, and one worldwide habitat. There is no place left for us to sail or run to.

Another famous example of human growth and collapse are the hunter gatherers of, the once naturally endowed, Fertile Crescent region, the great floodplain of the Tigris and Euphrates. Humans here evolved from this first human life style to the first human civilizations. Society such as Samarian, exploited the human ideology of agriculture and architecture and the effect of these human ideas are well recorded on the landscape. As human populations increased in the area more and more pressure was placed on the natural environment of the Fertile Crescent region. By 6,000 BCE, there is evidence of widespread deforestation and erosion of the area. The extensive practice of fire setting and overgrazing by goats might have been the chief cause of this. However, evidence shows that lime-burning, for the creation of plaster and whitewash, also destroyed the once vast woodlands. Today what is left is a thorny scrub and semi-desert world.

How can we stop this cycle of growth and collapse? There are many, many issues driving it: ideology, religion, politics, capitalism, democracy, and the most diabolical of all: human greed. When there was one billion pre-industrial humans on the Earth there were more than enough resources to give each and every one of them a good standard of living and still allow the Earth's eco-systems to flourish. However, now that there are over seven billion of us the Earth's resources are strained. When a natural eco-system is unstressed the creatures that share the system can live off the excess generated by it. Before industrialization and fossil fuel exploitation, we could live off of the interest that the Earth's eco-systems produced. Now we are consuming both the interest and the principal of all the Earth's eco-systems. Once the principal is spent there is no more interest, no more forests, no more rivers, no more air, and no more humans.

The next collapse humanity faces could be a worldwide one, epic in proportion, and perhaps the final one for the human species. However, if some of us do make it through this global downfall what will our descendants find in the geological record. Surely they will discover that the Anthropocene was a time of human dominance of the world, a time where humans held all the cards, a time when we could have stopped the cycle of growth and collapse, but a time that proved the human species was not intelligent enough to do so.

What is the next chapter in our geological history book now that humans have become a geological power, controlling the destiny of the only planet in the universe where life is known to exist?

What will ultimately happen to the human species? Will we survive or become extinct; as have 99.9% of all species that came before us? What will a future species see when they too read the sedimentary record? Will they discover a greedy, selfish, and shortsighted ancestor that thought themselves extremely intelligent, but were not smart enough to save themselves from extinction?

Perhaps, in the future our descendants will look back at the Anthropocene and rename it from: "the age of man" to: "the age of very, very, stupid men".

SOURCES

Scientists whose work contributed to the content of this book:

Ken Caldeira, Senior Climate Scientist at Carnegie Institute, California, USA
James Sevitsky, an earth scientist at the University of Colorado, Boulder, USA
Earl Ellis, an ecologist at the University of Maryland, Baltimore County, USA
Jim White, Professor of Geological Sciences at the University of Colorado, Boulder, USA
Jan Zalasiewicz, a geologist at the University of Leicester, U.K.
Gifford Miller, Professor of Geological Sciences at the Institute for Arctic and Alpine Research in Boulder, Colorado, USA
Baerbel Hoenisch, a marine biologist and geochemist at the department of Earth and Environmental Sciences of Columbia University, USA
Maureen Raymo, Director of the Lamont-Doherty Core Repository at Lamont-Doherty Earth Observatory of Columbia University, USA
Tony Barnosky, a paleontologist at the University of California, Berkeley, USA
Liz Hadley, a biologist at Stanford University in California, USA

ABOUT THE AUTHOR

About the Author

David Millett is a digital artist. He is an accomplished author, filmmaker, producer of paper books, and eBooks. He loves his wife Julia, writing, videography, photography, filmmaking, and travel.

Here is a list of his other works:

From David:

My Art: Book, Art

The Creature: Book, Fiction, Novel

Anthropocene: Book, Science, Environmental Science

Femlet: Screenplay, Drama, Shakespeare, Hamlet

Homo Cosmiens: Novel, Fiction, Sci-Fi, Drama

Autumn's Fall: Short Film, Drama

Table Talk: Short Film, Comedy

The Dog: Short Film, Drama

The Dialog: Short Film, Drama